Search Engine Testing –

Rating Web Search for Relevance and Quality

Author - Abhinav Vaid

Printed By Create Space – an Amazon Company

Available at **Amazon** and other retail outlets

Copyright

Copyright © 2014 by Abhinav Vaid

All rights reserved. No part of this publication may be reproduced, distributed, or transmitted in any form or by any means without the prior written permission of the publisher, except in the case of brief quotations embodied in critical reviews and certain other noncommercial uses permitted by copyright law. The web references, screen captures of web pages used in the current book come from publicly hosted web sites. The Products happen to be the property and ownership of the respective owners, and are referred to only for evaluation, research, and report findings. The findings are reported out of author's own research and study, and are intended for the masses, which could make choices that are more informed. The objective is education so that the audiences do not get trapped in the technological bandwagon. The author has permission of his employer for the use of his company name as an example. The other Organizational references are as per fair use under Section107 of Copyright Act. The information along with the research findings is shared purely for educational (noncommercial) purposes)

Table of Contents

About The Book 11
How to read the book/How the book is Organized? 16
A Career in Web Search 19
How Search Works 23
Search Engine - A Brief Primer 24
Classification of Rating 36
How Search Works 49
Getting Hired - A Career In Search 72
    Career Options in Microsoft Test Team 74

    Career Options in Google 76

    Career Options in Lion Bridge 78

    Career Options in Appen 81

    Career Options in Leapforce 83

    Social Media, Facebook and New Trends 84

Next Steps 85
FAQ'S 87
Play Safe – A Warning 89
References 93
Books 94
Work in Progress 95
About The Author 96
Disclaimer 97
Copyright 98
Feedback And Support 99
Acknowledgements 100
Blurb And Reviews 101

Author's Note To Readers 102
Contact 103

About The Book

Why This Book – Search engines that were initially introduced to serve as a reference standpoint eventually invaded human lives. As of date, they affect pretty much everyone in the world. The complete end-to-end process is either not known or happens to be incorrectly interpreted by many. The Search Engine Companies have published their best practices and recommendations in their respective Webmaster guidelines. This serves as one of the reference point for the players who work on web. Personally, I always thought that there is more to what it looks or what the search engine companies publish. It not only turned out to be true, but the people whom I spoke to also expressed their shock, because they always carried a wrong impression. Further research revealed that the employees who work on search engines could not reveal the facts because they are heavily guarded by trade secrets, confidentiality clauses, and NDA agreements. Another supporting factor is the history, which proves spammers exploiting it to get higher search engine positioning.

The downside of ignorance is losing the benefits of this completely new technological paradigm. One of the downside is the lack of awareness of career options, as they are typically not advertised.

The flexibility of timings with an added privilege of working from home bundled with a tag of working for a world class Organization makes it a lucrative career option. This kind of job does not demand high technical competencies. Anyone who has good command on language of choice (English and/ or Local Language), fair aptitude, and computer skills can consider this as a serious career option. Yes, decent salary, flexibility of working from home, and working for the best brand name. In a volatile industry, this can be one of the safest career options. A lot of effort has gone to keep the book short and crisp with an intention that the user should be able to start the book and complete it in a single read. Keeping it relevant and specific will cut down the lag time and help getting started immediately. The actual content is around 14 k words (around 50 pages) and accompanied with real world screen captures and flow charts.

Who this book is not for – This book is limited to the manual evaluation part, and does not cover the automation aspects of Search. The book also does not cover the algorithm that goes behind Panda. The book does not list the SEO rules, but after going through it, the reader will know with real-time examples what it takes to be there. This will provide answers to certain why's, which are not covered in hundreds of SEO books I have personally skimmed through.

This book is for - Usability Testers, UI Testers, and Web Testers who want to understand the manual testing and evaluation of Web Search. Anyone who wants to know the insights on how the web search works. The knowledge can be used to make informed choices for personal or professional tasks. It is highly recommended for people, who do not want to get trapped in the world of web. The live examples are accompanied with most recent real world examples of scams, detecting the spam and staying safe in the world of web.

1. **Career** - this book cannot guarantee a job, but will give a methodical direction towards –
2. **Doing the Groundwork** – Search Engine Introduction followed by concepts
3. Creating a winning resume that gets shortlisted
4. Types of Jobs available and Application process
5. Cracking the Exam
6. Building a Career in Web Search

7. Next Steps and Future

☐ In case the reader decides to consider it as a career option, the book will guide reaching the final gate in the shortest possible time. There will not be any surprises in the exam. The primary objective is that anyone with minimal knowledge of technology should be able to understand and reap the benefits.

Conventions Used

Refer to the table below for various conventions and their intended interpretation.

Convention	Interpretation
	Development - Technical References
	A bookmark
	A sign of Warning
	Web bot - Script to perform automation tasks
	Manual Rating/evaluation
	Future - What Lies Ahead
	Tip

Coverage – Due to the exhaustive scope of coverage, not all search engines could be tested. The findings reported in the current book come out of testing two search engines with respective percentages mentioned below –

1. **Google** - 100%
2. **Bing** - 50%

How to read the book/How the book is Organized?

The first part of the book starts with a high level Overview of Web Search and where it fits into overall life cycle. It is followed by complete manual process covering every type of rating. All the processes discussed are accompanied with supporting real world examples.

The second part of the book is about career in Search Engine companies like Google, Microsoft, and Lion Bridge etc. The knowledge gained in the part one will ensure that the applicant does not struggle during the qualifying exam. For the career aspirants interested only in the job opportunities can directly go to **Chapter Four** - Getting Hired - A Career in Web Search

Chapter One - Introduction

When an end user triggers a search, the Google crawler runs through indexes and delivers results that are meaningful to the end user. The results are delivered in a speed of less than one eighth of a second. There are billions of web pages in the background but the search results are tailored to user intent, location, relevance, and quality. There are very stringent rules and the parameters against which the relevance and quality of web pages is measured. Although there are teams writing the search algorithms and testing. However, there is another dimension to it. There are also highly skilled teams of Manual Testers, who have been working from the various parts of the world to make it a rich user experience. In case one goes through the Webmaster guidelines of Google or Bing, they do capture the best practices. What they do not have are the details of algorithms, and what is it that matters, with what percentage it matters, the locations (on the page), and the mathematical calculations of how the numbers are used. The most important part, as per my understanding, is the Manual Testing work that's being done at almost every nook and corner in the world. Although this kind of work creates a huge financial overhead, but then the users' gets rich quality, an attribute Google has always considered as a highest priority. This type of testing can be termed as a Web Based Usability Testing, Search Engine Testing, Usability Testing and so on. Additionally, there are a few more terms, which have evolved and the most common being "Google Rater", "Search Engine Evaluator", "Web Content Judge" and so on. For an overall understanding, it is mostly about the Organization for which the work is being done.

Organizations used different jargon due to business reasons, but the work, process, technology remains the same.

A Career in Web Search

One of the common, but wrong beliefs among job aspirants is the difficulty level in technical screening process. Although, the percentage to get shortlisted is very high and the time one needs to spend for preparation and appearing for exam can be taxing. Expected Skill Sets - The most important criteria are Language (you got it right), which is followed by aptitude skills, and then comes technical competencies. While technical skills can still be enhanced, but accuracy is the key and needs high level of concentration. What does it take to build a Career in Web Search? One week of solid preparation and you will reach the final gate without any roadblocks. A lot of effort has gone to keep the book short and crisp. A lot of flowcharts and real world examples are included to keep it relevant and engaging.

Business Value of Search Engine Testing - This particular segment has a lot of business value and worth investing time, considering what lies ahead. This is an industry, which will continue to grow for several years to follow.

Here are some numbers –

1. Two billion web users

2. More than 60 Trillion web pages

3. Web pages – a few billion get added every day

Tip – The staggering numbers is just a cursory snapshot. The data keeps increasing every second and so do the challenges of Search.

Tools and Prerequisites –

The infrastructural requirements of working for in any search engine company are more or less same, for example. Apart from above, there are some vendor specific requirements, for example.

1. **Microsoft** – Recommends latest IE Browser for Bing Testing
2. **Google** and Others – Recommend either FF or Chrome
3. **Specialized Platforms** – There is small work available for the native mobile platforms. This would require necessary mobile phone.
4. **Technology/domain specific** – There are jobs where high definition videos need to be evaluated. This would require high- end video cards.
5. **Other Factors** – A few additional factors need to be considered before taking a decision on career in Search Engine Testing.

Here is a high level list –

1. **Porn** – the sites can have the Porno material because there is no way the search engine company can predict that everything will be clean. Although, one can give unwillingness to work on Porno sites and those sites will not be sent to the evaluator.
2. **Manual Work can be monotonous** – I found the work to be monotonous, but that could just be me. There are people I know, who are fascinated to keep clicking the web pages one after the next. Something like this with a decent salary could be a rewarding experience to them.
3. **Importance of Accuracy** – This requires decent level of con- centration during the job. Since the payment is only for the quality and classification of content, the accuracy is important. When the accuracy goes below 90%, the red light will get

flagged and constant performance deterioration can go to the extent of losing the job.

4. **One Life and One Chance** - Some companies have stringent rules that allow only one attempt for the exam, during the entire lifetime. Yes, so the day you decide, make sure you do everything that it takes to knock the exam. It is certainly doable.

5. **Security** – It is not possible for the search engine company to evaluate the kinds of links pages have. Since the job requires clicking thousands of links, it exposes the individual to risks, which can be loss of data, virus infections, and so on. Although having a firewall, an anti- virus (with latest definitions) and On Access functionality enabled reduces the risk, but awareness is important.

6. **Technical Challenges** - It might not be frequent, but at times one does run into Technical Challenges that directly Influence the work. Being working remotely, it is quite natural to feel isolated and abandoned. Well, one should have some fall back mechanism in case one is not technically exposed. For the one is who don't know, there are hundreds of forums available online with people always willing to extend help. Do some homework to join the right forum. Although, the challenges are not complex and mostly confined to Operating System or browser related issues. In of issues that are typically seen and how to fix them. Additionally do not forget that you will be working with some of the smartest people in Indus- try. Most of the people I have known are genuine, down to earth, and helpful. They appreciate timely and clarity in communication. Salary,

Perks, and Other Employment benefits -

Salaries are paid on hourly basis and the amount is around $15 per hour. The amount can vary due to the location, level of experience, and skill sets. Since most of the jobs are work from home, there are no perks like medical insurance and other corporate benefits. At senior levels, there are people who have these kinds of benefits but that comes with lesser flexibilities like work from home. The work is typically around 22 hours per week, but can increase. The Organizations inform whenever that opportunity is available. Search Engine and Software Testing How is it different from Software Testing? Software Web Usability Testing is more or less driven by the project/client/product requirements. The A/B Testing is targeted towards specific goal for example marketing where they do it on two web pages in order to draw a comparison. The GUI standards are not in terms of black and white. In case there are, they get tailored based on other factors. The outcome of a Test Case is Pass or Fail. In Web Search Evaluation, the outcome is not Pass or Fail but categorization in terms of ratings based on the pre de- fined standards. The standards are common across the world and there is no scope of ambiguity, which is a common practice in Software Testing. In addition, it is not Testing the web pages. It is the Testing of Search effectiveness.

How Search Works

When an end user triggers a search, the results need to be qualitatively rich as well as related to intent of the query. If it were only mathematical algorithms making the query, it would result in endless number of results with unwarranted quality and relevance. For example, a farmer who lives in far-flung area needs some information on apples for farming. His query on apple will mostly give results of US based MNC, which he does not need.

Search Engine - A Brief Primer

At the highest level, there are 2 stages by which the data is made available to the end user –

1. Crawling
2. Indexing

The two terms have often been confused, although there is a clear-cut demarcation. It is important to understand that Crawling happens first, and then comes Indexing. Crawling is just like scanning the page, moving from link to link in a weaving like pattern. During crawling process, keywords are tracked, which later serve as an input for Indexing. Therefore, Indexing, in most cases, happens to be the bye product (but not always).

The Software Professionals with database background will find the concept similar to what they would have experienced during their career.

For non-techies, Google often refers it as a concept similar to a book's Index. Searching an Index (located at the end of the book) instead of searching through the entire book makes it quicker.

What Impacts Search? The two key factors that influence the search are relevance and quality. Whenever a search is triggered, a search engine has to –

1. Understand the location from where the query is triggered
2. Location from where the results are needed/ not needed
3. Language – localized or internationalized content
4. Relevance to user intent

5. Flagging the content type
6. Quality of search results (the quality of the search results displayed) -- Is there any search engine in the world that will be able to replace human intelligence to make the results so relevant, content rich, and specific? As of today, there isn't. Only some fractional portions are auto managed as the maturity continues to evolve.

Here is what happens –

1. Manual Testers/evaluators are assigned the task of rating the web sites by the search engine company.
2. They rate the web sites.
3. The results are submitted, which are re- verified and the pages classified based on ratings assigned.
4. There are certain smaller sized modular algorithms that are executed.
5. Pages are again evaluated/validated for the quality.

Comparison Analysis – A Comparison Analysis is drawn from the Results of Point 3 and Point 5

Objective – the approach serves two objectives –
1. The accurate results are taken into consideration and
2. The Effectiveness of algorithm gets validated.

Indexing plays a key role in speed by which the results are delivered.

Refer to the block diagram below, which highlights a high level process through which a data has to pass, before it is made available to the end user

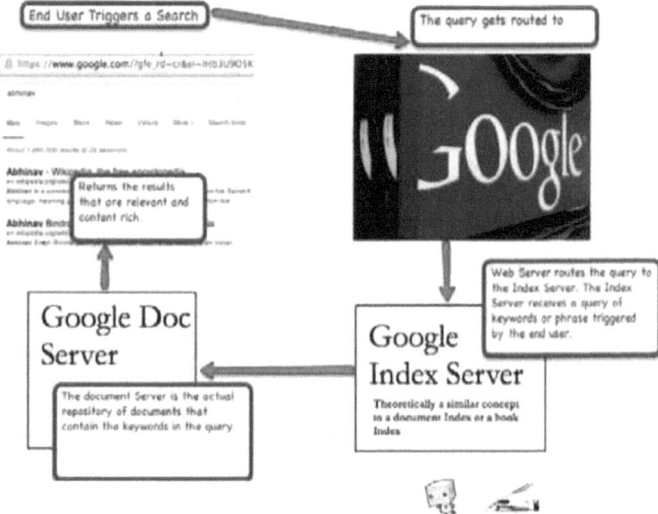

Figure 1 - How Google Search Works

Execution and Reporting - The first phase is to get feedback from evaluators, people who evaluate search quality based on our guidelines.

Regression Cycles – Multiple regression cycles are performed.

Quality Evaluation of Release - This is more of a Live or Release Testing where in the Evaluator is supposed to Test 2 versions of release. The first one is the Release X, and second one is Release Candidate.

Software Development Life Cycle of Search

The SDLC of Search Engine is more or less the same as other Software Products. Due to the dynamics of Web Search, Agile fits the best. Here are the high level processes that are typically followed –

1. **User Stories** – the User Stories typically include list of features or Change Requests for search improvement.
2. A typical data driven approach is followed which includes tweaking code and then Testing passing multiple sets of data.
3. **Quality Ratings** -- There are no Test Cases and the expectations are not Pass or Fail. The evaluators are required to specify the search results in terms of ratings. A guidelines document needs to be followed for Testing.
4. **Live/Hosting Server Testing** - Results are evaluated against the quality objectives and if they match, the code moves to live Testing.
5. **Release Certification** - The document is consolidated with all the necessary data and sent to the stakeholders for approval. If the CR is approved, the implementation follows.
6. **Web Scraping** - Web Scraping is a term often confused with Indexing, although it is closely related. Web scraping focuses more on the transformation of unstructured data.
7. **Keyword** - A keyword is the key phrase or the subject name. The two most important factors that play a key role in placement are with respect to Usage Location and frequency –
8. **Location** – page header, page headings and location triangle matters to most users and gets the highest ratings.
9. **Number** – refers to the number of times the keyword is dis- played in the web page. This is one of the traits exploited by spammers. This is also

referred as keyword stuffing, where the keywords are stuffed inside the page (multiple times) to the page to sound it relevant to the query.

During the later part of the book, there are real time examples that demonstrate how it was exploited by spammers. Google's spiders may also have some more advanced functions, such as being able to determine the difference between Web pages with actual content and redirect sites - pages that exist only to redirect traffic to a different Web page. Keyword placement plays a key role in how Google finds sites. Google looks for keywords throughout each Web page, but some sections are more important than others. Including the key- word in the Web page's title and headings is a good idea. Headings come in a variety of sizes, and keywords in larger headings are more valuable than in smaller headings. Keyword dispersal is also important. Webmaster should avoid overusing keywords, but many people recommend using them regularly throughout a page.

There are two areas, which typically every SEO expert focuses –

1. **Keyword Placement** - Physical location of the Keyword. **The Golden Triangle** – a typical term used by the SEO's to highlight the importance of Keyword placement. The statistical data proves it to be an area where most of the users decide, whether they want to continue to browse or move on to the next page. This makes it a key factor and the keywords in this particular area get higher points than others.

2. **Keyword Attributes /Properties** The Keywords in the Page Title, Headings (with Bold attribute), and Headers get the highest points. Adding the keywords in the Golden Triangle is the best-case scenario. A screen capture demonstrating the Golden Triangle is displayed

in Figure 2

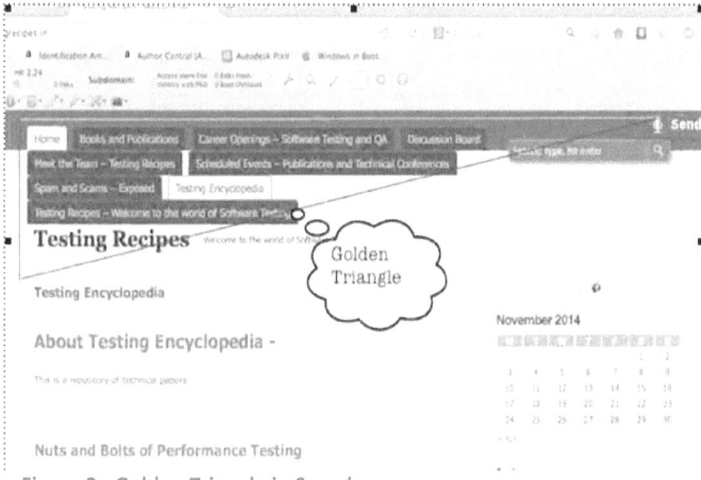

Figure 2 - Golden Triangle in Search

When I read it for the first time, it sounded too weird to be true. Curiosity prompted to dive deep and validate real-time findings.

At the time of compiling the current book, I came across the findings of two Organizations, who had published their eye tracking findings in a Press Release. A screen capture of the study findings is displayed in Figure Additionally, they also pointed out that There seems to be a "F" shaped scan pattern, where the eye tends to travel vertically along the far left side of the results looking for visual cues (relevant words, brands, etc.) and then scanning to the right if something caught the participant's attention. A copy of the Press Release is available at - http://www.prweb.com/releases/2005/03/prweb213516.htm

Figure 3 - Golden Triangle Eye Tracking Research

There is another search Firm – **Meditative**, which is dedicated into this kind of research, and had published its findings in 2005 but many of the findings hold true till this date. Refer to the screen capture of "Google Triangle Eye Tracking Research" in Figure 3. Google's eye tracking research report is available at - http://static.googleusercontent.com/media/research.google.com/hi//pubs/archive/34378.pdf

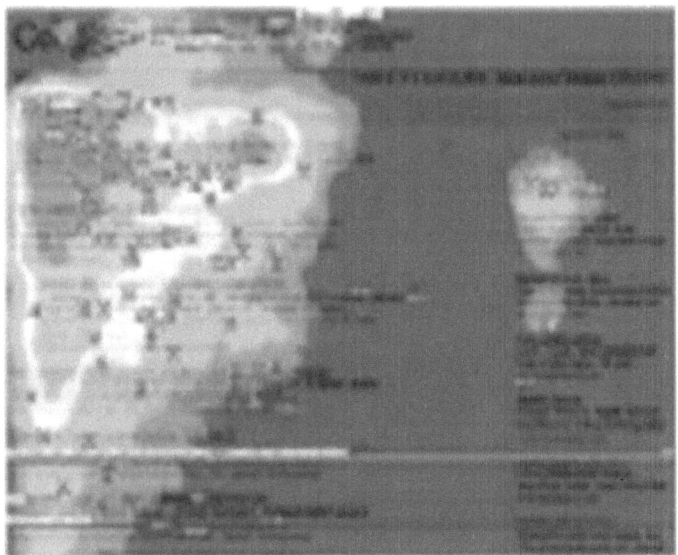

Figure 4 - Golden Triangle Heat Map

While there are tools and utilities available for tracking the traffic, clicks and so on, but there are tools which can go to the extent of diagnosing where the users make a cursory skimming of text and where a user actually spends time to read.

One of my favorite tools is Crazy Egg. They offer 90 days free trial as well, which is worth investing the time. The details can be checked out at - http://www.crazyegg.com/1stwebdesigner-90-days#

Relationship - Usability Study and Eye Tracking
Although there is a lot of research happening in and around Eye Tracking, which includes Head Tracking, Wearable Technology, Motion Tracking and so on. It is interesting to study the findings of eye tracking Web Usability study, which directly validates the points discussed. For further study, refer to - http://eyetrackingupdate.com/2010/06/14/eye-tracking-web- usability-study-reveals-golden-triangle.

Getting Familiar with Jargon - Understanding the Terms - Every technology comes with typical technical jargon and it is true in Web. Here are some terms, typically used in Search Engine Testing, and their intended meanings –

1. **SERP/Search Engine Results Page** – A result page that comes up once a user hits the search button.
2. **BOT** - The search script, which crawls the web to index results.
3. **MC** - refers to the Main Content
4. **SC** - refers to the Supplemental Content
5. **UR** - refers to Utility Rating
6. **QR** - refers to Web Quality Rating
7. **Intent** – any word can have multiple meanings.
8. For example, Arctern is an Indian Company, which is also the name of a US migratory bird. It is the users intent that de- fines what the user is expecting from the search
9. **Interpretation** – what the search engine under- stands from the query.
10. **NTRB** – refers to "No Title Results Block", when the SERP does not have a title. Refer to Figure 4.

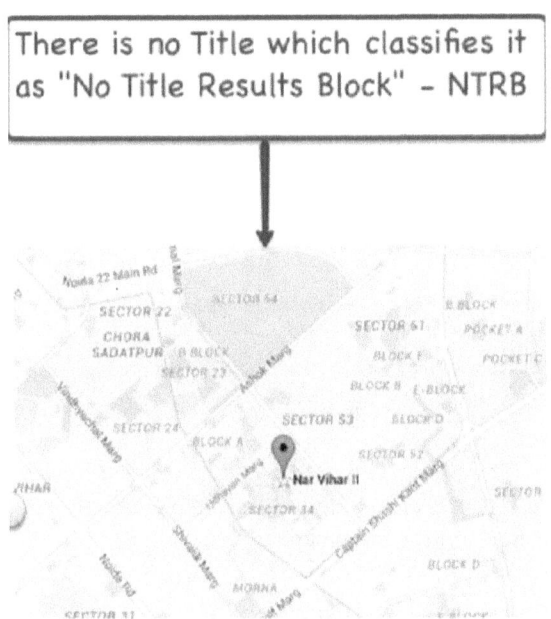

Figure 5 - No Title Results Block

11. **TLRB** – refers to "Title Link Results Block". When the Search Engine Results Page displays a Title based results, it becomes TLRB (refer to Figure 6)

Figure 6 Title Link Results Block

The Figure 6 displays a SERP, which has a Results based on Title. Here clicking the link results in directly navigating to what he end user is seeking (Figure7)

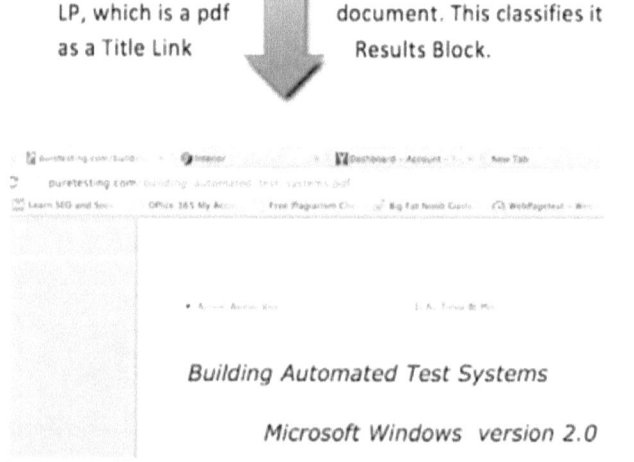

Figure 7 Landing Page TLRB

Classification of Rating

Types of Rating – at the highest level, there are two types of rating on the basis of which the web pages are rated. For clarity, please refer to block diagrammatic representation in Figure 7 that displays the high level ratings that need to be assigned to each and every search engine-rating task. The types of rating are

    1. **Utility Rating** – query dependent

2. **Page Quality Rating** – query independent

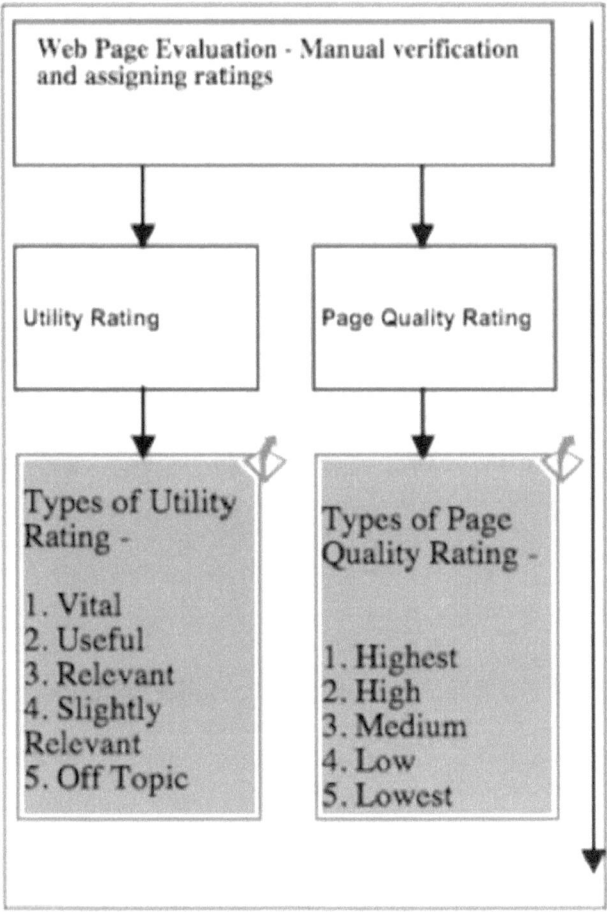

Figure 8 Types of Rating

Assigning Utility Rating - The following are the types of Utility Ratings –

1. **Vital** – refers to a clear-cut intent query for example Barrack Obama, and the result is exactly what the user wants

for example Barrack Oba- ma's Official Page. There can be other examples also, for example, a user looking for Microsoft and the result is Microsoft's Official web page. A clear-cut user intent, where the result is exactly what the end user is searching for.

2. **Useful** - a rating that is fractionally below the Vital typically comes under Useful. A useful rating is assigned whenever the results blocks are very helpful for most of the users.

3. **Relevant** - a relevant rating is assigned to the results that are helpful to many users or very helpful for some users. Relevant results have lesser valuable attributes as compared to Useful results. Relevant results still "fit" the query, but they might not be having the latest updates, come from a less authoritative source, or cover only one important aspect of the query.

4. **Slightly Relevant** – a web page missing to get into relevant goes into Slightly Relevant Category. Slightly Relevant may not have quality information, might be filled with stale information and so on.

5. **Off Topic** – an off topic or useless rating is assigned to results that are helpful to a very few users. Off topic results have a slight or no relevance to the subject and are full of keyword stuffing examples.

Utility Rating Scale – A Utility Rating Scale looks like this the one displayed in Figure 9. The ratings are classified in the order of importance with the least effective (Off Topic) being on left hand side and the most important(Vital) on the right hand side.

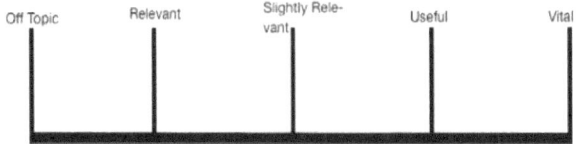

Figure 9 Utility Rating

Getting Familiar with Utility Rating - This refers to more or less the dictionary meaning of the term i.e., the usefulness of the web page. Utility Ratings are influenced by the query triggered by the end user.

Query – refers to the text or data entered in the search field User – refers to an end user, who enters text or data inside the search field

SERP – after a query is triggered, the page that shows up is known as Search Engine Results Page (SERP). The components that show up under SERP are known as Result Blocks and each block refers to one single result. For example, list search for the term Arctern in Google. Refer to the screen capture in Figure 10 with the results block screen. In the example, the page that is displayed is known as SERP. All the blocks lead to a different landing page. Notice that it is the name of a software company, which is located in India. The Author is in India, which is where he triggers the search, and the result shows based on location and intent.

Figure 10 Local Search Arctern

On the other hand, there is an American Bird with the same name, which does not show in the current search results.

Why? Since the bird does not stay in India, people do not know about it, which is why it does not show up.

Second scenario, what if the user wants to search a query in India and wants to know about bird and not about Organization? Solution – Change the search query to "Arctern Bird".

Or Reset the locale/settings of the browser The result will be different from the one displayed with local search.

Another Experiment – Lets do one more search, this time changing the browser settings to a country, which is foreign to both the Arctern Organization and the Arctern bird. A search in UK will give the results 360 degrees, the other way round - refer to Figure 10 Note - at times, based on the browser settings, one might be required to clear the cookies. Take a notice of how the intent of query can significantly change the Interpretation of result. Here the location plays a key role, for example in India; people would rarely search for bird. On the other hand, in

Western countries, the people searching for Arctern as an Organization will be relatively less.

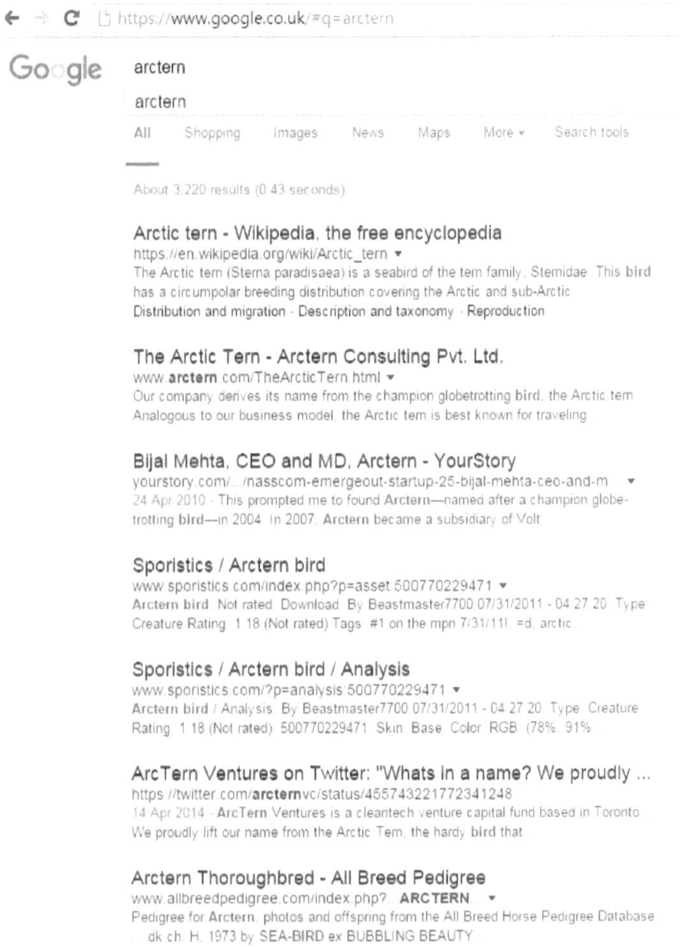

Figure 11 Foreign Language Search

Types of Interpretation - There are 3 types of Interpretation -

1. **Major Interpretation** – also known as dominant interpretation, it represents majority of the users

2. **Common Interpretation** – also known as dominant interpretation, it represents common users
3. **Minor Interpretation** – it represents some of the users

Classification of Intent – Intent is based on what the user expects to achieve from the result. There are three types of Intents –

Do – intends to perform an action, for example, taking a print out of bird

Know – intends for information for example, what does Arctern stands for

Go – intends to navigate for example, navigate to a Geological park, which has Arctern migratory bird in their park.

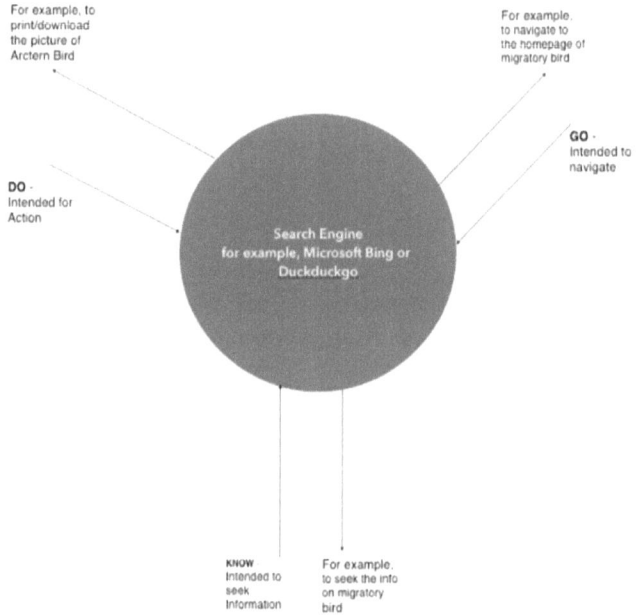

Figure 12 Intent Block Diagram

Another factor that comes into play is the local language. For example, people in Russia might not be looking for results in English and on the other side, a person who gets results in Russian but does not know the language will be of no help to him. Time factor is another criteria in Search. For some queries, the time of triggering the query influences the results where as for other cases, it does not. The Tester/Evaluator makes to make a judgment based on the scenario. If someone in India triggers a search for General Elections, the results block, which has no reference to 2014, will be of no use to the end user.

Web Page Structuring – How the Web Page is Organized?

A Web Page is typically structured into the following components –

1. **Message** - The core reason because of which the web page exists at the first place.
2. **Main Content** – Main message the web page is trying to communicate
3. **Supplemental Content** – supporting content to Main Content
4. Pages with a Helpful Purpose
5. **Content** - Understanding the Content
6. **Ads** – Advertisements on the web site
7. **Summary** – overall web page summary

Flagging Results - There are one more criteria known as Flagging, which does not influence the score of ratings, but needs to be reported for classification to different types. The different flags include Porn, Foreign Language, and Didn't Load Results. The results need to be flagged regardless of whether the query is seeking it or not. Porn flag needs to be assigned to all porn pages. A page has to be considered porn as long as it has pornographic content. The porn content can be anything from pictures, adds, text, videos and so on.

Foreign Language Flag - Foreign Language flag needs to be assigned when the language on the landing page is not one of the following: Default Language – can be English Local Language – can be English or the local language at the location where query is triggered.

Figure 13 Localized Language Results in Hindi

For example, most users in India use Hindi language. Therefore, the queries triggered from India resulting in landing pages in the Hindi language should not be flagged as Foreign. Refer to the screen capture in Figure 13. The flag needs to be assigned even if the Search Evaluator personally understands the language but most users in his location do not. All foreign pages need to be as- signed with the foreign flag, even if the query seeks a foreign language page.

User Location and Impact on Query - Every query has a locale. Some queries also have information about where the user is located. At times, the user location can change the interpretation of query. However, the statistics reveal that for most of the queries the user location from where the query is triggered usually does not change the interpretation. When is the user location important in under- standing query interpretation and user intent? As mentioned in the Arctern search example, one will need to use some common sense before setting the flag. One can also do some research in case there are doubts.

Queries with an Explicit Location

At times, the end users explicitly inform search engines about the kinds of results they are searching.

There are two ways to do location specific search –

a. Specifying the location in search preferences of browser.

b. By appending the target location name to the query.

At times, the explicit location might match the user location, and at times, it might not. -- Whenever there is an explicit location specified, the users expect the exact and accurate results from the search engine.

Vital Ratings for Queries with a User Location

In most of the cases, user location does not affect the vital rating. Rating Queries with User Location and Explicit Location The user location can play a large role in assigning utility ratings when -

The user location changes the understanding of the query or user intent. The query is a local intent query, i.e., the user is looking for local information or results nearby. For many queries, the user location is not very important when assigning utility ratings. The explicit location always plays a key role in understanding the query and assigning utility ratings.

Rating Local Intent Queries - When there is a user location for a local intent query, such as [Indian food] with a user location of Austin, TX, results in or near the user location are the most helpful.

Page Quality Evaluation

The second criteria for which Search Engines are evaluated are Page Quality. Just like Utility Rating, there is a slider here also for assigning the ratings. Getting familiar with Page Quality Concepts There can be lot of variations in Page Quality attributes. There can be high quality pages like organized, presentable, content rich etc. On the contrary, there are pages that are spammed, un- reliable, poorly organized, unhelpful, shallow, or even deceptive and malicious. These attributes are captured in Page Quality Ratings. Direct Impact One term that is very specific to this kind of Testing is when it comes to direct impact on individual's life. These kinds of pages are called Your Money Your Life. Any information that relates to the health or prosperity of an individual has to be important. Some typical examples include medical, physical, financial, etc. Since they directly influence the life of people, it is important that reputed publishers or organizations are endorsing the standards or the information presented.

How Search Works

When an end user triggers a search, the Google crawler runs through indexes and delivers results that are meaningful to the end user. The results are delivered in a speed of less than one eighth of a second. There are billions of web pages in the background but the search results are tailored to user intent, location, relevance, and quality. There are very stringent rules and the parameters against which the relevance and quality of web pages is measured. Although there are teams writing the search algorithms and testing. However, there is another dimension to it. There are also highly skilled teams of Manual Testers, who have been working from the various parts of the world to make it a rich user experience. In case one goes through the Webmaster guidelines of Google or Bing, they do capture the best practices. What they do not reveal are the details of algorithms, what matters, what percentage it matters, the locations (on the page), and the mathematical calculations of how the numbers are used. The most important part, as per my understanding, is the Manual Test Evaluation that's being done at almost every nook and corner in the world. Although it is expensive, but then it delivers rich user experience, something that's been Number 1 on Google's priority. This type of testing can be termed as a Web Based Usability Testing, Search Engine Testing, and so on.

INTRODUCTION TO SECURITY

There was a time when Security related challenges were restricted to Viruses, Trojan Horses, and so on. A trend that has significantly changed lately especially in the world of World Wide Web

Importance of Spam and Security in Web

Misleading and Deceitful Content

It is quite common to see web pages that display a separate content to the end users than one that goes into search engine indexing. They are kind of spoofed pages, which give a false projection to the end users but exist for some other reasons. One typical example is to make money-using ads or affiliate links. For example, some deceptive pages are designed to look as though they have helpful information, but in reality, they are created to get users to click on ads. It is recommended to report these pages even if one is not doing it as a part of the job. Google has been taking a lot of initiatives; refer to - http://www.google.co.in/insidesearch/howsearchworks/fighting-spam.html

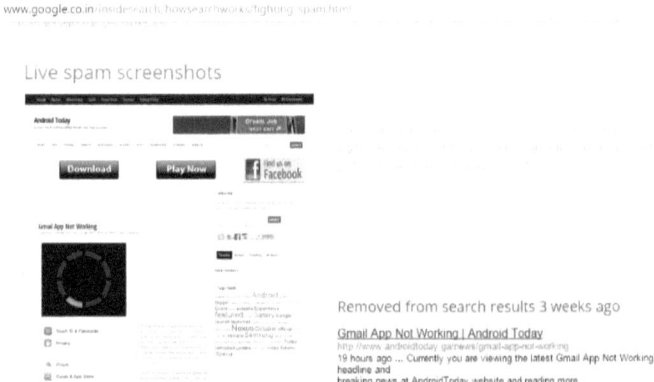

Figure 14 Example of Spam Site

Interested to know more, here are some numbers published by Google and their manual efforts to control the spam.

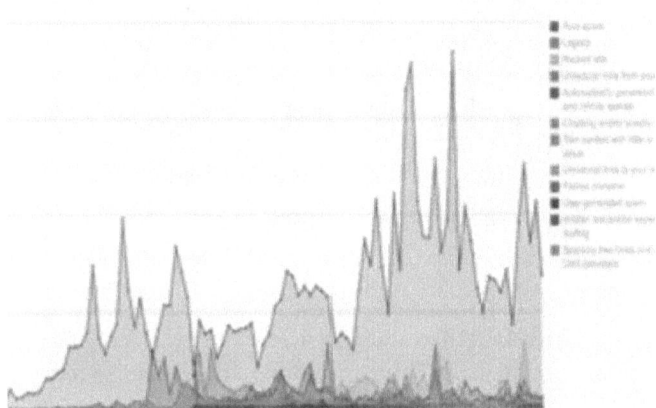

Figure 15 Spam Metrics

3. **PPC Pages** – Pay Per Click refers to the amount a site owner gets, once its ads are clicked. This has been exploited and there are multiple sites, which have nothing qualitative, and only serve the purpose of ads 4. Parked Domains – The domains are registered with no real purpose to serve the business, product etc. They can be used for a variety of tasks. In my personal experience, I once ended up registering a website with our name of our country included in URL. This I did on the suggestion of an expert in the web domain. Nothing unfortunate, except that the site name gave a feel of being hosted by Indian Government offering Telecom Services, which was completely wrong? On the other hand, the person had done all his homework. When the site came up with just one test page, it was generating a lot of traffic. Further diagnosis revealed that the site had many incoming links. It would have probably belonged to some Government entity, which might not have renewed and I ended up taking it. This particular business trait continues to exist till date. Recently I also read a bestseller, where the author had recommended checking the expired domains so that the new site owner could use the incoming links.

Where to draw the line?

There is no straight answer and one needs to use one's own judgment. In the above example also, the suggestion was not illegal. It was a personal gut feeling, due to which I dropped the owner- ship. After a gap of 12 years, I do see a rule today, which would have brought down its credibility.

1. **Keyword Stuffing** – refers to the number of times the keywords are repeated to increase the rating in Search Engine.

2. **Thin Affiliates** – refers to the site, which has very less useful content. For more information, refer to http://www.pickledshark.com/thin-affiliate-site

3. **Hidden Links** – nothing unfortunate about links, but Hidden Links are bad practices. In fact, even the paid links should be accompanied with a note with a disclosure statement should be explicitly. This is regardless of whether the links are visible or hidden.

There is another dimension to Hidden Links. The website can also be compromised by a hacker, and used to perform malicious activities. In that case, it is the website owner, or the Organization, which will get penalized. Refer to Figure 16, which is a classic example of this. It is interesting to know that the webmaster himself, technically competent enough did not realize till the time Google flagged it.

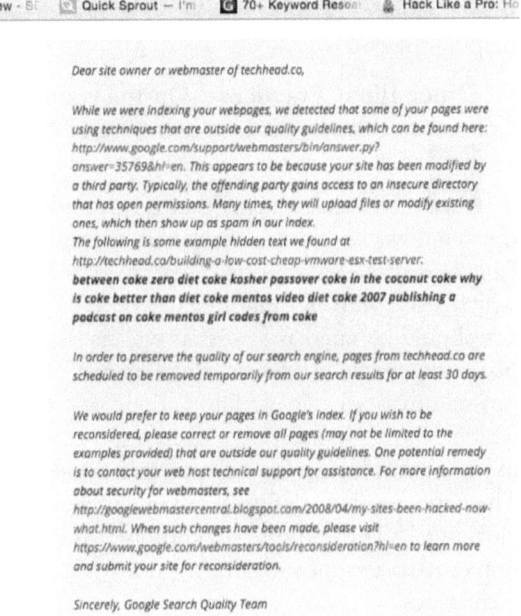

Figure 16 Hidden Links

4. **Sneaky Redirects** – refers to redirect a visitor to a different URL, then what was initially requested.

Attributes of Highest Quantity

Role of Main Content – Generally, as a rule of thumb, the highest quality pages are well crafted, professional, and supported with original and authoritative content. They are known to provide richest experience to end- users Role of Supplemental Content - At times, Supplemental Content can also contribute towards High quality. For ex- ample, if you are searching for Usability Testing Online Tips, and you get to a Usability Testing Article, or a book, then it is indeed helping the end user.

Other Illicit Practices - On the contrary, Supplemental Content can also be distracting or spam to the end user.

– The spam pages that do not allow a user to close the window, or closing results in multiple windows getting launched should always be categorized as Lowest. The pages, which exist with nothing qualitative and are just serving for ads should be rated as lowest as well as flagged for spam. It is not very easy to detect these kinds of issues, at least programmatically. In addition, there are thousands of sites, which have redirection operators configured and the moment you login, you will get redirected to a different site. The term used to define this kind of behavior is known as cloaking. The most common and relatively easy implementation of cloaking is done via the following two techniques –

a. **Multi frame** – This is the easiest technique, which pretty much anyone can implement. In order to implement this, two frames are used and one frame covers the entire visible area of the screen. There is no way any web user will be able to notice this. On the other hand, the web crawler will be able to see both the frames.

b. **JS** – Java scripts are generally used for redirection. The script navigates an end user to a different page as com- pared to a search engine. That is another reason, why some of the high security browsers disable java script as a default behavior.

Examples of PPC - There can be extreme examples of PPC, where in there will not be any meaningful content except the links.

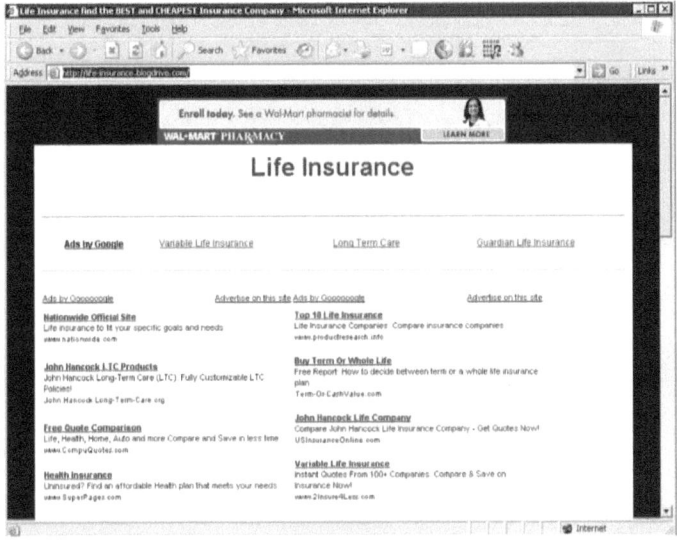

Figure 17 Page with No Meaningful Content

It is always recommended not to click the links that look deceptive, nor share email ids on non-reputed websites. I have personally seen the monetary numbers that these people get by following dubious business traits. Spammers, regardless of what they claim are a menace to the world of web and we as users, should do what- ever we can to make it a healthier and cleaner environment. For additional examples, please refer to http://research.microsoft.com/enus/um/redmond/projects/strider/searchdefender/GoogleSpam_Examples.htm for Google's examples.

Homepage Checks – It is not possible to check each and every page for the maintenance of content but homepage cannot be ignored. One needs to ensure that at least the homepage is updated. In order to do this, perform the following steps – Check the validity of copyright dates

Contact information – the contact information on the

web sites need to be available and should be accurate. This is especially true when the relationship is with respect to medicines, credit cards, or some critical product support involving financial transactions. Any dubious information should be reported. For example, a web site having online support has a broken contact page and wrong contact details can be immediately flagged. Please refer to the following example. This is a world reputed Organization with hundreds of positive paid reviews, but most of the functionalities the devices claim to support are broken. They have not mentioned the contact number. There is an SMS number, which does not seem to work. The only way to reach the vendor, which is not possible because the functionality of contact form is broken.

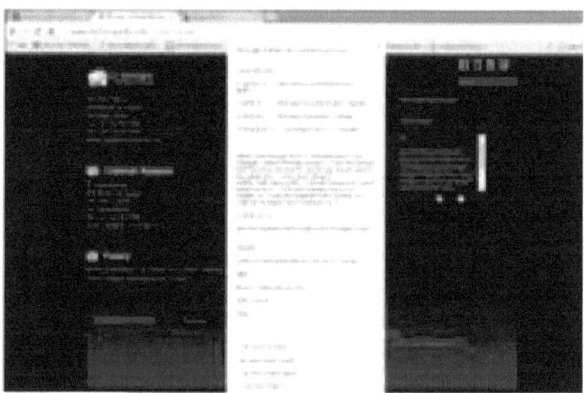

Figure 18 Contact Us Broken Functionality

Fake Reviews - do not get trapped - it is quite common to get trapped due to fake reviews. Here are some live examples - The example in Figure 18 is a typical example of getting trapped due to fake reviews. I purchased this phone after going through recommendations that came from highly reputed channels and newspapers. The press and media channels do not have resources to personally validate everything, and heavily depend on the reviews and that is how I got duped. I went to the extent of reaching out to the Organizations that had published the reviews, but no one ever acknowledged as the reviews were manipulated. During the publication of the current book, various agents, individuals, and Organizations approached me offering various services. I got stumped because I not in touch with anyone and completely focused on book. A few minutes of research revealed that the publication data as well as interactions with publisher were pulled out with the help of scripts. I was promised certain services, which could push the book to the top of Amazon's charts. New York Times exposed this particular trait.

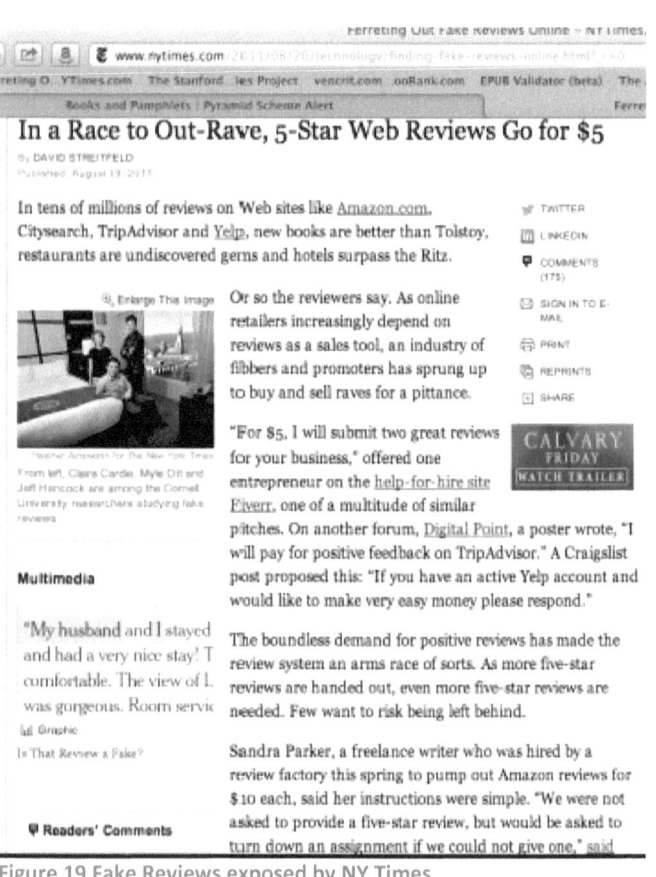

Figure 19 Fake Reviews exposed by NY Times

Ever noticed astronomical numbers in Facebook likes, huge number of tweeter followers, sudden raise in 5 star ratings for the crappy products that are not even ready to ship. Now you have the answer.

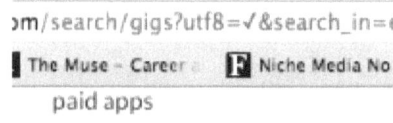

Figure 20 Mobile App Review in $5

Mobile App Review - This gentleman is from India, offering 25 Reviews for any paid app at the cost of $5

Who are Fake Reviewers - human beings like us who proudly exhibit their competencies to perform deceitful services for a fee of $5. Please refer to the following two examples -

Book Reviews - do not buy books purely on the basis of review feedbacks. I have a collection of over 200 books, and on some occasions, the books turned out to be completely different as com- pared to what was projected in 5 star reviews. Being in IT profession, I was always short of time and that is why I never paid a close attention to anything except review feedbacks. Its only now I realize being duped because re- views were falsely projected. I recently discovered a book where in the Author had changed his first name to the name of the book and then booked a website containing the same name. Everything aligned perfectly, followed by paid reviews. No surprises because he was destined to make it to first page of Google. Please refer to Figure 21 that displays the kindle book re- view service offered at the cost of $5

Figure 21 Kindle Review in $5

Solution? There is no straightforward solution and can vary based on situation. When in doubt, cross checks the credentials of everyone (the vendor as well as the reviewer). I always ensure to alert the platform that is being exploited, which in this particular case is Amazon

Other Dubious Traits and Illicit Practices

Truth about Expired domains - when an individual or an Organization drops (no longer ex- tends) the subscription of domain, it is called an expired domain. One of the key factor to deter- mine the monetary value of any website is the quality of links. By registering an expired do- main, the new site owner gets the benefit of preexisting links. It is quite common to come across web marketers using expired domains.

Do Not get trapped. A little research on the domain name, the services offered by the domain, and the reputation of site owner would reveal the facts.

Plagiarism – There are millions of cases on web, where people take the text from other sites, reword it and publish it by their own name. There are also cases of open source text or PLR (a license where the new owner pays to publish the article with its own name). any similar practices should be flagged and reported. One good site to check for plagiarism is http://www.paperrater.com/plagiarism checker

Content Mills – There was a time when there used to be thousands of Con- tent Mills on the web, where in the authors would get money to write articles. The money varied from $1 to $20. The articles obviously used to have little, if any value to the readers. It became so hit that some tools also got released. One would only need to enter some keywords, enter some parameters like relevance, quality, research, length and tool would create the article. The author would get money for doing practically nothing, at the cost of web users time. A few months back, Google came up heavily on these sites and penalized/ delisted them. There are sites that still exist but they have raised their standards of quality. When in doubt, use the Plagiarism check. Guru's claim that the article should give an impression that no one in this whole world, except the author could have written it.

Reputation - The reputation of a website is also very important when the information on the website demands a high level of authoritativeness or expertise, such as medical information web- sites. When a high level of authoritativeness or expertise is needed, the reputation of a website should be judged by what expert opinions have to say. For example, marketing or non-technical people endorsing the latest mobile phone that got launched. Alternatively, a software engineer citing the medical web site.

Maintained/updated website – Election News, Medical, legal, financial, and other Your Money Your Life related web sites need to remain up- dated.

Encyclopedia References - There are many encyclopedia-type websites. Some are highly respected publications appearing in all major libraries and are standard references. Some are websites edited by anonymous users with no editorial oversight or fact checking. Reputation research is extremely important. High and Highest page quality ratings can only be used for encyclopedias with very good reputations. Wikipedia articles with a lot of detailed, information-rich Main Content and external references can usually be rated in the High range. In rare cases, Wikipedia articles may even be rated as Highest. Image Checks – there have been thousands of cases with respect to illegitimate Images used by people, who do not have the appropriate rights to. This however is not an area of concern for the evaluator, but knowledge is important.

When in doubt, check the authenticity of image. This can be done by going to URL - http://ideeinc.com/

If there is no Main Content on the page, its rating should always be lowest. Page quality ratings are query-independent - The rating depends on how well the page achieves its purpose and how well it passes page and website checks.

If page or website falls short of quality standards; the page should always be given a low rating. A high rating needs to be given only if the web page, summary, title, and message are in synch.

Google's Guidelines and Recommendations – Although the guidelines are confidential, a 160 pages guidelines document was leaked out by a "General Guidelines Version 3.27 – June 22, 2012" Just make a search in Google with the text high- lighted in Bold, and you will have the document. The same document is given to the candidate in order to prepare for the qualifying exam.

We talked about Google standards, what about Microsoft?

Microsoft follows a little different jargon than Google. Having said that, the intended meanings and the interpretation remain the same. Microsoft has always been known for user-friendly ness, and they have applied similar concept in search engine classification as well. Microsoft guidelines require color codes to classify the ratings. One cursory look is all it takes to get it right. Here are the terms that Microsoft uses, along with their equivalent rating in Google

Perfect - Official page with most likely intent. This is similar to Vital in Google.

Excellent - This is similar to Useful in Google. Bing describes this as a landing page that "strongly satisfies a and "closely matches the requirements of the query in scope, freshness, authority, market and language."

Good – Equivalent of Relevant in Google. A Good landing page "moderately satisfies a very likely or most

likely intent, or strongly satisfies a likely intent." Bing says most searchers would not be completely satisfied with one of these pages and would continue searching. According to Microsoft, a Good page should address the intent of at least 10 percent of searchers.

Fair - Equivalent of Slightly Relevant in Google. This rating applies to pages that are only useful to some searchers. A Fair page "weakly satisfies a very likely or most likely intent, moderately satisfies a likely intent, or strongly satisfies an unlikely intent." A Fair page addresses the intent of at least one percent of searchers.

Bad - Equivalent to OT in Google. Here there is one addition done by Microsoft. The spam pages, parked domains and redirected sites are also need to be classified as Bad.

Strong	Moderate	Weak	Poor	Obscene Content	Content not accessible
Excellent	Good	Fair	Bad	Detrimental	No Comments
Excellent	Good	Fair	Bad	Detrimental	No Comments
Fair	Bad	Bad	Bad	Detrimental	No Comments
Bad	Bad	Bad	Bad	Detrimental	No Comments
Bad	Bad	Bad	Bad	Detrimental	No Comments

Likeliness of Intent

Figure 22 Microsoft Standards and Ratings

Final Words - Quality speaks for itself - After going through all the aspects, rules, standards etc., and the bottom line still remains the same i.e., Quality speaks for itself. Humanly judgment is always knows over and above the standards for example, one can go to some web sites and a cursory glimpse itself shows where it stands. The supporting examples can be www.apple.com, www.google.com, http://www.britannica.com/ and so on. Do validate all the sites mentioned above and cross check the standards mentioned in this book. There will not be a single best practice missed on any of the areas discussed.

Reporting and Task Submission

As discussed in the first chapter, this particular work does not follow traditional Software Life Cycle Models including exhaustive reporting. In fact, the task reporting is crisp and to the point, with no scope of ambiguity, whatsoever.

The actual task assignment form is displayed in Figure 23

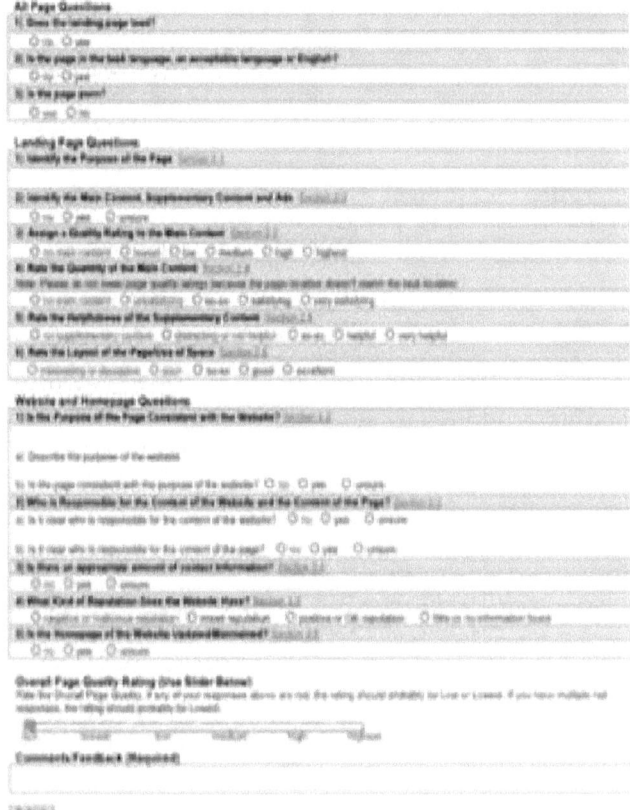

Figure 23 Sample Task Submission

In the Software Industry, there was a time when aptitude was one of the key skills being validated, followed by language/ technology skills, and finally domain. For example, the Organizations working in Banking or Financial domain would prefer candidates only from similar domains and similar is the case with other streams like security, Infrastructure, and so on. The similar trends have started picking up in Search Engine Testing also. Jobs have started being advertised for specific area or domain. Since it is infancy stage, it is easier to get into a platform of choice to have a strong foundation. Some of the popular areas include Advertisements, Video, Usability, Spam and Security. Another area, as discussed earlier is going to be Language specific jobs.

Other Technical Considerations – There are some technical factors that impact Search Engine related testing. A high level over- view of most common impacts is mentioned be- low -

Mobile Devices - Mobile has a huge penetration and the world is migrating very quickly. Typically, the jobs with respect to Mobility are separately advertised. In case they are not, the testing is pretty much on the html 5 sites. This in turn does not have any significant impact on the way the testing is done. In case the testing is specific to mobile segment, there are many factors that will need to be considered right from connectivity, GSM/GPRS, Bluetooth, CDMA, WLAN (with a combination of various standards like A, B, G, and N). Then comes the Operating System for example, an- droid, iOS, Windows Phone and finally browser type and version. Each of these factors will impact how the testing will be required to be per- formed, which is beyond the scope of the book. Web page rendering depends on variety of factors and none of the points mentioned above can be discounted.

How are the above factors supposed to influence the web search? The expectations of the end users change. It was referred it Search Engines take fraction of a second to deliver the results. Mobile segment takes more time, as the users are more tolerant. Each of the connectivity options mentioned above impacts the search results. As a ballpark standard estimate, a GPS/GPRS based cellphone takes around 8 seconds to load a page. However, the results would change if the connectivity gets changed to Wireless LAN. In Wireless LAN also, the results would vary if one were connected via Open Mode or any other Security Option. Testing Solution - The various vendors have released tools/apps that simulate the environment of the targeted mobile device. For example, I often used to test the application on http://ipadpeek.com, which simulates a similar UI Interface. There are also some plug-ins available, which can be directly installed on the browser to simulate the targeted mobile device.

Browser Specific Issues - At times, there are issues with respect to browser types and related functionality, which makes it a challenge in terms of how to handle it. The first step should be tried to recreate it in a different browser. In case the functionality is working fine, then it is almost certain that the issue is browser specific. Research is the best way to handle it. For example, Google Chrome has a unique feature known as WebGL. The technology enables viewing extreme content for example 3D. It is at times very common to see the content not loading or even browser crash. Since the implementation is relatively new, it will take time to understand the issue, which can be due to multiple reasons. One typical error message that browser throws is WebGL has failed: "Rats! WebGL hit a snag..."The more information you attach while reporting these kinds of bugs, the more it will help the software vendor. Workarounds in most of the is- sues are more or less the same –

1. Restart the browser
2. Reset the browser settings
3. Un-install and re-install the browser.
4. Restart the system

Role of Operating Systems - At a theoretical level, it is easy to assume that any issue/problem/bug that occurs would be with Browser regardless of the base Operating System. Unfortunately, that is not the case. For example, x version of Internet Explorer on a Windows XP or Windows 2012 data centric server or MAC would not produce the same results. When in doubt, instead of flagging it directly, it might be worth attempting to browser version/type.

Browser versions, Updates, and Add-ons - The expectations and requirements are very clearly communicated by the Organizations. It is important to ensure that you have the right browser type, version, updates (if any), and add-ons recommended. The accuracy of results can have huge variations in results.

Embedded devices - The alternate embedded devices have every- thing different than a traditional laptop/desk- top. For example, cellphones have ARM based processor and stripped down versions of browsers. The devices are seriously con- strained/limited resources and can never ever give the same experience that one would get in a laptop or a desktop. The difference can be from version x device to the version y. Regardless of what the vendors claim, it is al- ways worth cross checking it against alternative device and browser. The most common factor is the heavy consumption of heap by the mobile devices. Regardless of what the vendors claim, the challenges continue to exist till date. Al- though due to the recent advancements and cutting edge processors, the number of cases is reducing with time. Testing in any device would always produce variations in results, when it is retested on emulator or alternative device. Re- search and make an informed choice.

Getting Hired - A Career In Search

The hiring process is highly intensive. After you submit your application, you will be sent a document, which has the training material on the basis of which you need to prepare for the exam. The key to success is read page to page at least a couple of times. Go through each and every instruction carefully because they are derived out of deep research. Any fall through might result in failure. For ex- ample, one of the key search engine leaders gives one week for completion of 3 exams. It recommends appearing for the first exam within 2 days, followed by 2nd on 4th day and 3rd on the last/final day. The preparation time recommended is 10 hours and each exam duration can stretch over 10 hours. That means, if one wants to give all 3 exams on the last day then it will be highly taxing on mind as well as body. The key here is accuracy, which one typically loses to maintain for very long periods. One of the key players have a knock out criteria according to which, one can appear for this exam only once during the entire lifetime. So there is no scope of committing a mistake. Here is a list of Organizations, application URL's and the application process –

Crafting A Resume That Gets Shortlisted

The first step is submission of a resume that gets shortlisted. Although, I have not seen very stringent resume scanning, but there are some recommendations to make it foolproof.

1. **Readability** – any typos, grammatical errors, spelling mistakes are going to ensure that the resume goes to trash. Check for readability and make sure that it is crisp and to the point.

2. **Accuracy** – whatever you present should be accurate and measureable, the numbers, data, and your character of what you really are. It is okay to be fair and present the picture of what you are instead of what you aspire to be. These Organizations are very professional and appreciate straightforward and clear-cut communication.

3. **Demonstration of Expertise** – demonstrate that you have certain knowledge of web as a whole. Doing a small write-up of whatever you know best, and mentioning the link in resume will take you a long way.

4. **Core Technical Roles** – in case you do not want to limit yourself only into manual and want to get into more technical roles, then have a good command on algorithms, problems related to pattern matching, regular expressions, text/data extraction etc. A fair knowledge of "data mining", scripting language like Perl, and a Programming language like "C" will help.

Career Options in Microsoft Test Team

Software Test Engineer (SDET) II Job Category: Software Engineering: Development Location: Vancouver, CA Job ID: 865945-133715 Division: (Not Group Specific) Do you have strong skill set or desire to learn the following: Search, Relevance, Machine Learning, or Natural Language Processing? SQL Server Business Intelligence is a strategic and rapidly grow- ing business for Microsoft and one of the fastest growing data technologies businesses in the industry. From complex data processing to a host of rich visualizations, we are right now in the Golden Age for Microsoft BI. We have a super exciting and rare opportunity for you to have a huge impact in helping ship amazingly cool world- class quality software solutions to customers where two extremely hot technology areas converge - business intelligence and natural language processing. The Power Q&A team needs talented test engineers to build and test the next generation of business intelligence software for mil- lions of Microsoft customers. We are an innovative startup team that blends the exciting worlds of natural language processing and data visualization. The team is growing fast, our ship schedule is ambitious, and there are major challenges ahead of us. We are passionate about being the first to deliver an experience that enables everyone to navigate the world's data using rich visualizations and the latest natural user interface technologies. If you share that passion, you are a great candidate for the team! On our team, you will have the chance to work on a V1 product with huge business impact and Microsoft-wide scope. We are releasing to SharePoint/ O365, and will leverage the latest NLP technology and data resources from Microsoft Research and Bing. In this position, you will work on relevance measurements for the NLP engine for Power Q&A. You'll read the latest research papers, collaborate with other NLP and machine learning developers across Microsoft, build and test fast/shippable runtimes. Ultimately, you will find ways to get the latest NLP technologies out of the lab and into the hands of customers.
Responsibilities - Understand customer scenarios - Design metrics using NLP techniques and understanding, data sets using structured and

unstructured data, to accurately mea- sure the quality of the product and these scenarios. - Using C# and T-SQL, design and implement infrastructure and tools to improve quality at scale - Provide technical leadership in designing and implementing test systems across BI Test Team - Collaborate with developers and program managers to deliver high quality product, on time - Initiate and promote engineering best practices, and contribute to engineering excellence initiatives in collaboration with peer teams - Fill gaps and innovate infrastructure and tools written in C# and T-SQL to improve engineering productivities

Qualification - 5+ years of systems software development experience and preferably in a test-engineering role - Strong C/C++/C# programming background - Using metrics as a means to measure quality - Able to make critical decisions independently in a fast paced and demanding environment - Prior experience on BI or working on natural language processing, Search, or Relevance is a definite plus but not absolutely required - A BS/MS degree in Computer Science or related technical field is preferred.

Career Options in Google
Ads Quality Rating

Google still has some people in its team who do similar kind of work. This particular opening is directly with Google but on contractual.. The URL to the Google career site, which has the position details, is embedded below -
https://www.google.com/about/careers/search

Ads Quality Rater (Temporary). English Language UK Specific
Mountain View, CA, USA

Quality Evaluator (Temporary)
Los Angeles, CA, USA

Ads Quality Rater (Temporary). Spanish Language
Mountain View, CA, USA

Ads Quality Rater (Temporary). Russian Language
Mountain View, CA, USA

Figure 24 Career Options in Google - Ads Quality Rating

For the Job Description, Refer to Figure 25

v.google.com/about/careers/search/#!t=jo&jid=43087& — Ads Quality Rater (Temporary), English

chnologies - Solutions With Care... Ads Qu

Ads Quality Rater (Temporary), English

Mountain View, CA, USA

Product & Customer Support · Temporary

Know someone who would be interested?

Find connections Sign in to see your connections at Google

For immediate consideration, please send a text (ASCII) or HTML version of your resume to temporaryjobs@google.com.

Important: The subject field of your email must include **Ads Quality Rater (Temporary), English**.

This is a temporary role offered through ZeroChaos, Inc., formerly WorkforceLogic.

ZeroChaos, Inc. is recruiting part-time telecommuters with fluency in English and an in depth and up-t date familiarity with English-speaking web culture and media to help with Quality Evaluation for websites for Google Inc., the search engine company based in Mountain View, California.

As an Ads Quality Rater, you will be responsible for reporting and tracking the visual quality and content accuracy of Google advertisements. Ads Quality Raters use an online tool to examine advertising-related data of different kinds and provide feedback and analysis to Google. Projects worked on may involve examining and analyzing text, web pages, images, and other kinds of information. You will need an in depth and up-to-date familiarity with English-speaking web culture anc media. Additionally, you will apply this knowledge to a broad range of interests and topics. Ads Qualit Raters possess excellent written communication skills and web analytic capabilities. You will be required to work 10-30 hours a week on a self-directed schedule. A secure, private high speed interne connection is required.

Responsibilities

- Evaluate the accuracy of Google web advertising.
- Communicate effectiveness of web layouts and information via the online tool.

Minimum qualifications

- BA/BS or equivalent degree or equivalent

Preferred qualifications

- Excellent analytical capabilities.

Figure 25 Job Description Google

Career Options in Lion Bridge

Google used to have in house dedicated teams to do Search Engine work, a trend that changed during recent years. Most of the work is being done by Lion Bridge. They have a lot of openings, at the time of writing this book. They are professionally strong and have decent reputation. Navigating to the Lion Bridge site displays 2 openings i.e., for Internet Accessor, and Internet Judge. Refer to the screen captures below.

Word of Caution – Apply only and only if you are 100% sure to give the exam. They give one chance during the entire lifetime and in case you fail once, you will not get second chance.

The exams get scheduled on Wednesday and you are given the study material on the basis of which you need to clear three exams. It is recommended that you appear for the first exam on 2nd day, second exam on 4th day, and the third one on the last day. Each exam is a knock out exam and in case you fail in once, you cannot go the next. The passing percentage is 90%. It had sounded very easy to me, but it was very taxing due to the exam durations.

Ref – What We Offer Flexible Hours. You will have the flexibility and freedom to work from your own home, working your own hours, Sunday – Saturday, depending on availability of tasks Further opportunities may arise to contribute to other tasks and projects on a freelance basis High Speed Internet connection. Minimum up- load speed of 1mpbs Software applications, e.g. Antivirus software, Adobe Flash Player, Acrobat Reader, Adobe Shockwave player, Microsoft Silverlight etc. Technical Requirements Work from home office environment

PC with Windows 7 or Vista. Apple OS or older versions of Windows are not acceptable Flexibility to complete a variety of different tasks following standard sets of guidelines

Strong attention to detail, analytical skills and excellent communication are essential Ability to work independently and flexibly to new techniques/ processes Degree level qualification Requirements Fluency in English is essential, additional language skills are an advantage Must be working and living in India for a period of 5 years Familiarity with current and historical business, media, sport, news, social media and cultural affairs etc. in your country A keen interest in the Internet e.g. researcher, blog writing, forums, website publishing At the time of writing this book, there are three variety of openings available in Lion Bridge India. It might be interesting to notice that there are a variety of choices one can opt for when it comes to the type of work you aspire to do. For example, the work also has other domains such as multimedia, languages and so on. Refer to the Figure 26 –

India

Web Content Judge – India `English`

Lionbridge is the leading provider of translation, development and testing solutions that enable clients to create, release, manage maintain their technology applications and Web content globally. Through our dedicated Enterprise Crowdsourcing division, we h clients grow their businesses by providing the most agile and professional work solutions on the planet changing the way work is done. To learn more visit http://www.thesmartcrowd.com

Internet Assessor – India `English`

Help play a part in improving the quality of one of the largest search engines in the world. For this role we are looking for candid who are based in India and are fluent in English.

Multimedia Judge – India (English Language) `English`

Help contribute towards making internet search and online images & videos more exciting, relevant and interesting for all end us your market. We are looking for candidates who are based in India and who are fluent in English.

Figure 26 Career Options in Lion Bridge

Career Options in Appen

Appen Butler Hill also enjoys a decent reputation and has huge list of openings. At the time of writing this book, there are around 1400 jobs available (globally). Refer to Figure 27

Figure 27 1400 Jobs - Appen

Featured Jobs with Appen - Appen presents this jobs

in the Featured job bank. Refer to Figure 28

Figure 28 Featured Openings Appen

Refer to Figure 25, which displays jobs for Search Engine Evaluator in multiple geographical locations

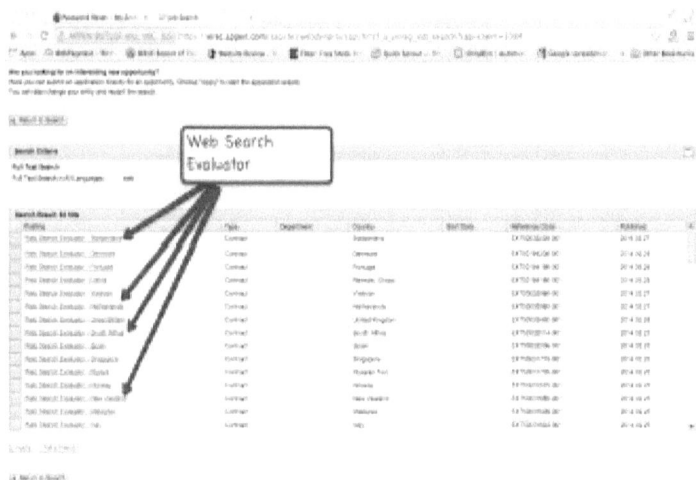

Figure 29 Web Search Evaluator - Appen

Career Options in Leapforce

The application process in Leapforce is more or less on the same lines of other Organizations. Refer to Figure 30 for details –

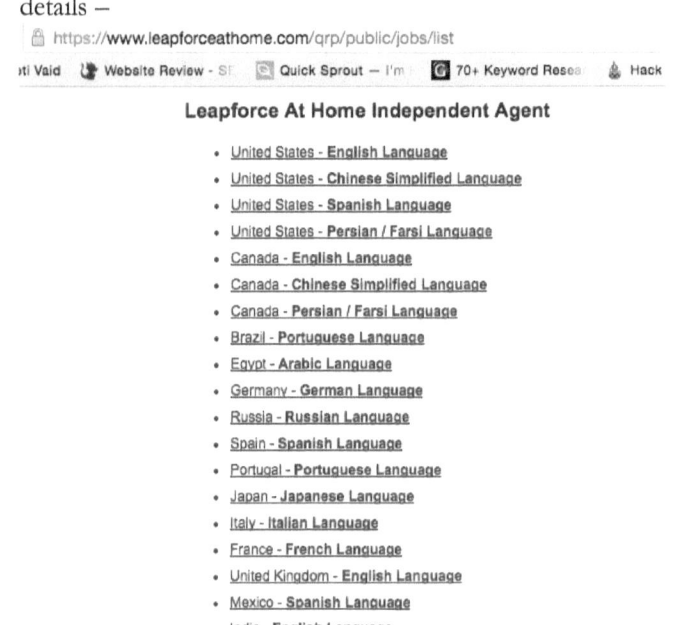

Figure 30 Openings in Leapforce

Social Media, Facebook and New Trends

Initially the work was limited only to Search Engine Evaluation but that Social Media explosion has taken a different direction. Currently Facebook is hiring for the rating jobs that show on the Facebook news feed. The work is again being offered via various third party companies that include Leapforce and Appen Butler Hill.

In case interested to work via Leapforce, refer to the screen capture with the link to help you register for the program.

> ✓ Due to the nature of my confidentiality agreements with the vendor, I cannot go into specifics. The current screen capture does give clarity on the transparency with which these Organizations work.

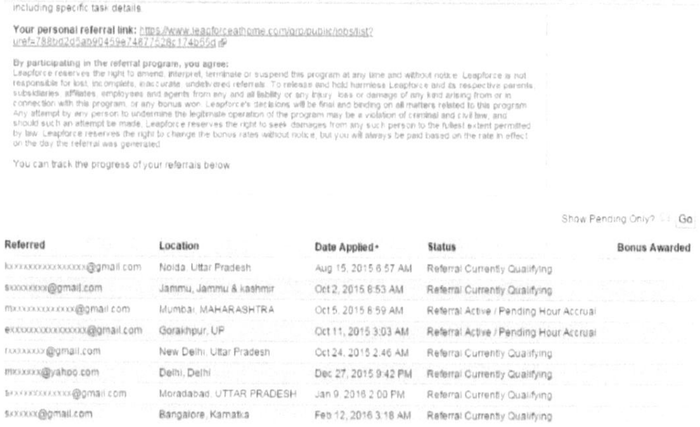

Figure 31 Social Media - Facebook Program

Next Steps

Decide if you want to take it as a profession. In case you do, here are the sequential steps -

1. It is recommended to Setup a PayPal account. Although there are few players, who do the wire transfer but it largely depends on the location/ country.
2. Apply for job. You will get a buffer time to clear the exam. Prepare well. Reach out to author in case there are still any doubts.
3. Go through each and every word of exam instructions mail. It is important to have accurate understanding of exam pattern. Also, go through the Organizational recommendations for appearing in exam.
4. Appear for the Exam. After the exam, the results are immediately displayed.
5. Start getting the work

Research

In case one really wants to go that extra mile on Research, here are the options –

1. It might be worth putting together documenting 210 rules, which do exist, but are nowhere documented.
2. Go through server logs and figure out if Google's team visited your site. Interesting!
3. Artificial Intelligence. Some of the major IT giants including Microsoft. Amazon, and Google have been making efforts towards incorporating Artificial Intelligence. It might be worth investing time for this lucrative area of work. Amazon has already built a product on similar lines. http://searchenginewatch.com/article/2328015/The- Evolution-of-Artificial-Intelligence-in-Search- An-SEO- Opportunity
4. Creating your own Search Engine

FAQ'S

Why are not jobs outsourced to the companies that are specialized for delivering off shore work? This is one of the most frequent questions often being asked. There is no straight answer because there are plenty of reasons. The process is confidential and there will remain a scope of information being leaked out. Most of the companies that work on the onsite – offshore model has a reputation of not delivering on time, lack of quality, and un-realistic billing. For example, one person working and billing done for multiple people or inflating the number of hours. Additionally, there would always be a risk of information being leaked out.

Where can we look for additional Information? There are some references available, which can be referred to via Google search. The best source would be connecting to someone at Google's Team or with Microsoft Bing. In case not successful, please feel free to reach out to me and I will get back within one working day.

Localization and future work? Every language in this world needs to be rated and categorized apart from English, be it Hindi, French, Chinese, Japanese, or even Tamil. There is some work done in few popular languages but most of the work is pending, the volume of which keeps piling up every second. There will be a lot of work coming up for these kinds of tasks and the hiring process might be less rigorous considering the limited number of resources available for local languages.

Platform Coverage and Scope – The book has been created using MAC. The work has been verified in all the popular browsers including Fire- fox, Chrome, and Internet Explorer. The Operating Systems used for the work include Windows 7 and MAC. For readability and flow, the digital testing of supported eBook formats has been tested on multiple mobile emulators as well as on Micromax Canvas Doodle 2 (ARM/An- droid 4.2)

Play Safe – A Warning

There is one problem in the world of web, the process is almost irreversible. Feel free to skip this chapter, in case not interested. Sharing some experiences that only highlight how ugly things can become, in case one is not careful –

H.263 – It is a high quality video format used in mobile phones. I was responsible for making the release, which we managed after slogging days and nights in office. It was an innovation, destined to change how the videos were captured, transmitted, and viewed. It got released after getting integrated with the latest mobile phone by world's number one cell- phone company. One of the video taken in that format by a student went viral. Result, the Managing Director of largest online store got arrested.

YouTube – There is a video available on YouTube by my name. It shows a birthday par- ty being celebrated with alcohol. I am nowhere to be seen in the video, but I see a few people who look familiar and happen to be from my industry. They are drinking, making fun of IT Industry, bosses, salaries, and so on. After a long back and forth with YouTube, I stopped chasing them to get the video removed. A woman I know is facing a divorce, because her information about her affair was shared with her husband by a popular search engine Height of exploitation - Please go through this video http://www.youtube.com/watch?v=Z0LZ6DN CgrY and do some research before you enter the world of web. It is estimated that at least 30% of all computers are infected with malware.

Ignorance can be Suicidal - Working on the web has another issue associated with awareness. The laws governing the Cyber Space are relatively and lack of awareness has further added to confusion. There have been cases, where the lack in awareness led to extreme. Whenever in doubt, check the law before proceeding.

Some real life incidents –

A former colleague, who used to report to me, uploaded software modified by him to a non- profit Open Source Portal. He was terminated, his profile was blacklisted. Another incident, a person working for Apple shared something casually, for a not so significant product. He could not bear the consequences of the treatment he received and eventually committed suicide.

When in doubt, either - Check the legal aspects Don't visit the website if you are suspicious or, Check the authenticity of website url before visiting the site - http://onlinelinkscan.com"

Figure 1 - How Google Search Works 26
Figure 2 - Golden Triangle in Search.................................29
Figure 3 - Golden Triangle Eye Tracking Research............30
Figure 4 - Golden Triangle Heat Map 31
Figure 5 - No Title Results Block 33
Figure 6 Title Link Results Block...34
Figure 7 Landing Page TLRB...35
Figure 8 Types of Rating 37
Figure 9 Utility Rating 39
Figure 10 Local Search Arctern 40
Figure 11 Foreign Language Search 42
Figure 12 Intent Block Diagram 44
Figure 13 Localized Language Results in Hindi 46
Figure 14 Example of Spam Site 50
Figure 15 Spam Metrics..51
Figure 16 Hidden Links 53
Figure 17 Page with No Meaningful Content 55
Figure 18 Contact Us Broken Functionality 56
Figure 19 Fake Reviews exposed by NY Times 58
Figure 20 Mobile App Review in $5...................................59
Figure 21 Kindle Review in $5..60
Figure 22 Microsoft Standards and Ratings......................65
Figure 23 Sample Task Submission 67
Figure 24 Career Options in Google - Ads Quality Rating.76
Figure 25 Job Description Google 77
Figure 26 Career Options in Lion Bridge 80

Figure 27 1400 Jobs - Appen 81
Figure 28 Featured Openings Appen 82
Figure 29 Web Search Evaluator - Appen 82
Figure 30 Openings in Leapforce 83
Figure 31 Social Media - Facebook Program 84

References

http://www.testandtry.com/2009/04/20/7-ways-to-quickly-rate-website-quality

https://www.ets.org/erater/about/faq/

http://www.trabalhodigital.com/search-engine-evaluator.htm

http://www.microsoft-careers.com/key/bing-jobs- Ads-Quality-Rater.html

http://searchengineland.com/library/google/google-search-quality-raters

http://searchengineland.com/google-gutted-its- search-quality-rating-guidelines-for-public- release-150281

http://searchengineland.com/bing-search- quality-rating-guidelines-130592

http://www.ncbi.nlm.nih.gov/pubmed/18760843

http://searchengineland.com/interview-google- search-quality-rater-108702

http://www.web-rater.com/tags/test

http://www.seobythesea.com/2011/11/how-human-evaluators-might-help-decide-upon-rankings-for-search-results-at-google

http://yahoob- ingnetwork.com/en-sea/home

http://www.bruceclay.com/blog/bing-powerful-new-search-choice/

http://searchengineland.com/google-search-quality-raters-instructions-gain-new-page-quality-guidelines- 132436

http://www.nytimes.com/2012/03/01/technology/impatient-web-users-flee-slow-loading-sites.html?pagewanted=all&_r=0

Books

If there were books available on this subject, the current book would not have been written. Although, there are some recommended books which can be used as an indirect reference.

1. Search Engine Optimization: Your visual blueprint for effective Internet marketing by Kristopher B. Jones

2. SEO Fitness Workbook: The Seven Steps to Search Engine Optimization Success on Google by Jason McDonald

3. Google's PageRank and Beyond - The Science of Search Engine By Amy Langville (Author), Carl Meyer (Author)

4. The Ultimate SEO Bible By James Mason

Work in Progress

Bringing Intelligence in Search Automation - The research is in progress with a skeletal framework in place. The project is about incorporating automation to replace manual efforts by equally intelligent scripts.

The goal is to achieve increased coverage across multiple platforms, without using any commercial tool.

About The Author

Abhinav is an IT professional and has spent 15 years in the Industry working for various blue chip companies including Motorola, McAfee, Lotus Interworks, and Arctern.

Some of the key products he has released include Java cross compiler Tool Chains for MIPS, Code Warrior Tool Chains, McAfee GSE, and ConsolVMS. He is Foundation member of ISTQB Indian Board and an associate of Computer Society of India. He is the author of "Building Automated Test Systems", and has been writing for various Technical and Research Journals.

Disclaimer

The entire text in the current book comes out of author's own research and study. The work has nothing to do with his current and/or previous employers. The references to the Public domain are cited. There are some examples where the author has used the name of his employer for which he has the permission. The findings in the current book are author's own interpretation of the web search. This by any means is not intended to challenge any existing standards. Use it as per your judgment and understanding as the results can vary based on variety of reasons. The author and the publisher may not be held accountable for any damage caused due to the findings reported in the book.

Copyright

Copyright © 2014 by Abhinav Vaid All rights reserved.

No part of this publication may be re- produced, distributed, or transmitted in any form or by any means without the prior written permission of the publisher, except in the case of brief quotations embodied in critical reviews and certain other noncommercial uses permitted by copyright law. The web references, screen captures of web pages used in the current book come from publicly hosted web sites. The Products happen to be the property and ownership of the respective owners, and are referred to only for evaluation, research, and report findings. The findings are reported out of my own research and study, and are intended for the masses, which could make choices that are more informed. The objective is education so that the audiences do not get trapped in the technological bandwagon. The author has permission of his employer for the use of his company name as an example. The other Organizational references are as per fair use under Section 107 of Copyright Act. The information along with the research findings is shared purely for educational (non- commercial) purposes.

Feedback And Support

Feedback and Support - Any Technical questions, doubts, feedbacks, and scope of improvements should be addressed to the author. Unless there is high level of complexities, all queries will get addressed within one business day.

Live Webinar Events - In case the readers want a live demonstration of the work reported in the current book, please feel free to contact the author along with expectations, time zone, and preferred date to host the event.

abhinav@testingrecipes.in

Acknowledgements

This book would not have been published without the support of multiple people. A note of thanks to Mr. Surya Katakam, Vice President Arctern/Volt for Organizational permissions. My Industry colleagues Manmohan Pandey (Arctern Consulting), Vikram Reddy (Arctern Consulting), Vipul Kocher (ITB, Salt), Vivek Sharma (Oracle Corp), Meenakshi (QA Lead Sopra), Rishi Raj Koul (Head of Network Operations, NSN), and Shiv Kumar (Sr. Manager, Documentation, Freescale) for their feedbacks and reviews.

A special note of thanks to - Shiv Kumar, Sr. Documentation Manager at Freescale and Rishi Raj Koul, India head of NSN for their quick turn- around times for blurbs. Manmohan Pandey and Vikram Reddy, who spent endless number of hours working on reviewing the contents.

Blurb And Reviews

Shiv Kumar Sr. Technical Documentation Man-ager at Freescale Semiconductor, Inc. Abhinav Vaid is professionally known to me and has good command in the field of Software Testing. In the past several years he has built a good knowledge base around software testing that has enabled him to implement out-of-the-box testing methodologies at work. His coworkers respect him for his attention to detail (ATD) and dive deep nature. I have reviewed his current Book on Search Engines and I feel he has done justice to the subject.

Rishi Raj Koul Head of Network Performance – Nokia Siemens Networks India & Africa I have known Abhinav from his Motorola days. As in person, and always, he continues with his trademark earnestness and straight-forwardness even in the narrative of his new book about Internet search engines. I find the book very lucid and full of knowledge that will help budding IT aspirants to wade through the quagmire of IT jargon and labyrinth of protocols. Be it an Engineering student or a budding Testing professional, this book pro-vides invaluable information that is not avail-able in a single compilation anywhere to my knowledge. I wish Abhinav and the readers of this book all the success.

Author's Note To Readers

Why I wrote this book?

In the world of cyber space, it is quite common to see people trapped on the wrong side of the fence, with no clues how they got there and where to look for solution. On the other side, as a part of the IT Industry, it is quite common to witness Organizations exploiting tools/ technologies because the customers are not in- formed enough to make the right choices. The laws governing the web space are either not clear or still in the process of evolution. Companies like Google and Microsoft have been on their toes making every possible effort to control the menace, but the people on the other side also have some smart techies who take pride in negating the efforts. Additionally, the career options in this particular segment offer a lot of flexibilities and decent work life balance. My take - this is the first time I have done a simple, non-technical job considering the impact it can have. Having said, multiple people have reviewed it and all of them have strong technical background. Therefore, I am sure there be a scope of improvement in the manner in which is information is presented. Please share any feedback that could improve the quality of the book. Thank you for investing valuable time reading the book. In case you need any help, any clarifications on the information presented, please feel free to get in touch.

Abhinav Vaid

Contact

Mail-abhinav@testingrecipes.in

Skype – abhinavvaid

Voicemail www.testingrecipes.in

Other Coordinates –

Facebook - facebook.com/testingrecipes

Blog - searchenginetesting.blogspot.com

YouTube - https://www.youtube.com/watch?v=9U2O0yuC0xk

Twitter - @vaidabhinav1

Amazon - amazon.com/author/abhinavvaid

End of the Book

#################

www.ingramcontent.com/pod-product-compliance
Lightning Source LLC
Chambersburg PA
CBHW030906180526
45163CB00004B/1724